Climate Guardians

Thoughts and Actions for People of Faith

David Cain

Emily,
Thank you for
this lovely
conversation.
Keep the Faith!
Dave Cain

Climate Guardians

Thoughts and Actions
for People of Faith

David Cain

Montemayor Press
Montpelier, Vermont

Photo credit: NASA

Printed in the United States of America

For information contact:

Montemayor Press
P. O. Box 546
Montpelier, VT 05601

Web site: www.MontemayorPress.com

1 3 5 7 9 10 8 6 4 2

To my five grandchildren
in hopes of sustaining a livable world:
Carter, Desmond, William, Owen, and Tim

Contents

Introduction i

Part I - THOUGHTS v

1 - The Promise of Climate Stewardship 1

2 - An Unoriginal Sin. 5

3 - Church of Carbon 11

4 - Leap of Faith 17

5 - Climate Awakening 23

6 - "God Will Save Us" 29

7 - The Climate Bystander Syndrome. 35

Part II - ACTIONS 39

Notes 43

Climate Guardians

Thoughts and Actions for People of Faith

David Cain

Introduction

This little book is the byproduct of my efforts to uncover the reality of climate change. No, not scientific reality, but the *full* reality. The book is an attempt to rebalance forces that Pope Francis, in his Encyclical "*Laudato Si:* On Care for Our Common Home," warns are preventing us from learning "how to live wisely, to think deeply and to love generously."

First, a confession. I'm confronting a climate change paradox. I believe that this global situation is very real and very dangerous. But I confess that like everyday climate change deniers, I have a fossil fuel dependency. Religions teach us that we're all sinners. That label fits me—and it's humbling.

My denial paradox is strangely analogous to one of the most famous modern scientific discoveries, the "uncertainty principle." When one attempts to measure one of two different but related physical properties, determining one of the properties with greater precision results in greater uncertainty in the other.

Yes, climate science and climate morality are related. Yet as science has gotten closer and closer to pinning down the reality of human-caused climate change and its consequences, our moral certitude has frayed—at least if we gauge the situation by noting our individual and collective inaction. Individually and collectively, we can't stand the implications. Agnosticism and apathy are at play here.

Why? Many of us won't be personally impacted by climate change to any significant degree. Who, then?

Other people in different places and generations will suffer the direst consequences. We who are living in the early twenty-first century may take incremental steps to reduce our carbon footprints. We're empathetic. But empathy simply isn't good enough, and we know it.

This modest collection of essays attempts to resolve that uncertainty by sketching out a moral arc. Each essay serves as a navigation point in thought. The first essay considers God's grant of dominion and human proprietorship. That proprietorship isn't earthly domination, but stewardship. The next essay explores the hypothesis that climate change, like slavery, is not original sin and therefore not automatically subject to His grace.

The trajectory of the moral arc bends to the horizon with an essay that compares the structure of an oil drilling platform to the layers of passive iniquity that fossil fuels play in our lives. At the arc's apogee is an essay about a leap in faith, describing how individual climate commitment is supported and safeguarded by the canopy of all the major world religions.

The glide path along the moral arc is marked in a subsequent essay about spiritual awakening and personal climate-change witnessing.

The final two essays seek to bring climate morality down to earth—dismissing, first of all, the notion that God will save us; and, second, imploring us to act as Good Samaritans against pluralistic ignorance. Here is where the concepts of good stewardship (at the trajectory's beginning) and good Samaritans (at the trajectory's end) are intended to infuse thought with action. That is where I believe full climate reality lies.

I realize that secular readers and those of other religious traditions may have difficulty embracing the notions of "climate sin and salvation." Regardless of your religious beliefs, however, it's my fervent hope that by adding your own thought narratives to this moral trajectory, we can strengthen it. In doing so, we resolve the uncertainty paradox all together as we move beyond climate empathy toward action. That action will cause climate altruism to grow exponentially among us, summoning the better angels of our natures. This is the essence of climate guardianship.

Part I

THOUGHTS

1

The Promise of Climate Stewardship

A February 15, 2015, article in *The Deseret News,* a publication owned by the Church of Latter Day Saints (the Mormons) is titled: "Dominion & Stewardship—The Disappearance of Dominion Thinking in the Christian Climate Change Debate." Giving this topic a full-page spread in its Faith section is significant in its own right, underscoring the notion that climate change is a matter of grave spiritual concern.[1]

Climate change, and our role in causing it, are among the great moral issues of our time. Many religions have adopted positions urging members to confront these issues, but the Mormon linkage of climate and stewardship drives right to our common moral core.

There has been a marked shift in public attitude (since the 1960's) about humankind's dominion over nature. According to a 2014 Public Religion Research Institute study, 57% of U.S. adults believe that humans should be "stewards" of the earth's creatures and creation, while only 35% believe humans favor "dominion."[2] This represents a sea change in western attitudes since the start of the Enlightenment period hundreds of years earlier.

Why is this so? The traditional pretext is in Genesis 1:28, in which God exhorts man to "Be fruitful, and multiply, and replenish the earth, and subdue it: and have dominion over the fish of the sea, and over the fowl of the air, and over every living thing that moveth upon the earth."[3] The word "dominion" has justified the mindset that man can use the earth's resources for his own ends and as he sees fit. That includes unfettered extraction and burning immense quantities of fossil fuels, as well as exploiting the earth in many other ways as a grant of dominion by God.

But *The Deseret News* article points out that "dominion" in English versions of the Bible is translated from the Hebrew word *radah*. It also notes a common problem in translation where languages do not align: direct word substitutions without carrying forward the source's underlying context often convey the wrong meaning. In this case, the context for *radah* in the original text is "benevolent rule." Good *radah* is increasingly interpreted by biblical scholars as "stewardship." Michael Stead's Anglican essay, "To 'Rule Over' and 'Subdue' the Creation," exemplifies this evolution in religious thinking. Stewardship is also a key element of Pope Francis's 2015 encyclical on climate change, "*Laudato Si:* On Care for Our Common Home."

Growing popular support for "stewardship" over "dominion" during the past 50 years may actually stem from that other word in Genesis 1:28: "subdue." If there were ever a time in recorded history when humans have managed to subdue nature, at least temporarily, this would be it. In many parts of today's world we live in "gardens" controlled and shaped by our own technology. Ordinary people unconsciously sense the increasing fragility of nature under assault from human activities. And they are commonly associating environmental stewardship with virtuous behavior. This conjoining of religious teachings and the popular stewardship has great promise for broad environmental awakening and push for climate stewardship.

2

An Unoriginal Sin

The politician attacks climate change with gusto in his Christian youth conference address.[1] "The Climate is changing and climate has always changed and always will," he intones. "[M]y point is, God's still up there. The arrogance of people to think that we, human beings, would be able to change what He is doing in the climate is to me outrageous."

Is this politician committing a sin by implying climate change is God's will, and is he absolving himself and the youths in his audience of any responsibility?

Talking "sin" is like touching morality's "third rail" in this day and age. Nevertheless, given the gravity of the challenges we face from climate change, a breach of protocol is warranted. What follows is a lay person's perspective.[2]

Many years ago, humans struggled with the sin of slavery in the same way that we're struggling with climate change today. There is no explicit prohibition in the Bible against slavery. There is universal agreement today, however, that condemns the practice of slavery as sinful.

In his book *White Over Black,* historian Winthrop Jordan writes that "Many Americans awoke to the fact that a hitherto unquestioned social institution [slavery] has spread its roots not only throughout the economic structure of much of the country but into their minds."

The divide between proslavery and antislavery constituencies in the U.S. was fierce. People working against slavery often put their lives on the line. Religious institutions were at the forefront.

But many people and institutions tolerated slavery, overlooking its cruelty and manifold moral contradictions. After all, a slave economy can generate immense wealth and prosperity for those who aren't slaves.

One apologist for slavery at the time conceded that slavery was a sin, but it was an original sin—in other words, not a sin you can do anything about.[3] Because ending slavery would bring ruination upon slaveowners and slaves alike, the greater sin would be emancipation.

What tipped the balance were people who faced up to the sin of slavery. They acted. They broke through the casual acceptance and apathy. They ultimately fought a civil war. Slaves were emancipated. The practice of slavery became a sin in the minds of the religious and secular alike.

With climate change, we have arrived at a similar situation. Like slavery, there is no biblical prohibition against carbon fuels being dug out of the ground and burned.

But to paraphrase slavery historian Jordan, "a hitherto unquestioned use of fossil fuels has spread its roots not only throughout our economic structure of much of our country but into our minds." Many are saying that curtailing carbon fuel use will bring economic ruination.

So, let's return to the statements of the politician quoted at the start of this essay. His thesis is that God is changing the climate, not we ourselves, and that it's arrogant for any of us to assume otherwise.

This politician is effectively maintaining that carbon pollution is an original sin. Since God's hands are at the controls, there's nothing humans can do about climate change, and we're therefore absolved of any consequences. All humans are doing is exercising His grant of dominion, to prosper and multiply as God intended.

Not so fast!

When God created Earth, He included a natural carbon cycle that maintains the Earth's temperature. Natural mechanisms maintain the carbon dioxide level in the atmosphere within a benign range. God put in a thermostat and got Himself out of hands-on temperature management.

Some would say it's a miracle that carbon dioxide concentration is perfectly balanced within the range that supports life, while buffering variations in the planet's orbit and occasional volcanic eruptions. People of faith would see this not as a miracle, but part of God's creation.

Therefore, is it an "unoriginal" sin to continue reliance on carbon fuels, distorting God's carbon cycle and wrecking His thermostat? The rationalizations that we're now hearing as we see undeniable signs of climate change are of the same order as those used by slavery's apologists.

Against this backdrop the following questions are manifest:

Is willful ignorance about (climate change or slavery) a sin?

Is acknowledgement that human activities (burning fossil fuels or condoning slavery) have distasteful and sometimes tragic consequences justifiable for the reason of their leading to greater economic good? But are they not still sinful?

Is absolution warranted when one feels trapped and

pacified by prevailing political mores and economic circumstances (fossil fuel or slave-powered economy)?

Each person must decide. For many, the words of a blogger stand out:

"I'd have reservations using the word 'sin' in describing environmental ethics myself, mainly because the term is so easily misunderstood . . . Nevertheless, there is a changing 'sense of wrongdoing in the wider society,' so it amounts to the same thing, really, whichever word you choose for it. We can identify as 'sin' those things that we previously did not recognize to be so. The time is coming when it will be hard to call yourself a Christian and live an earth-destructive lifestyle." [4]

3

Church of Carbon

Alighieri is a well-known travel writer and sometimes poet. This is one of his series of articles about the architecture of human creations found around the world. In this article, Alighieri recounts his recent architectural tour of an offshore oil platform—truly our era's Church of Carbon.

As our boat approached the black, hulking edifice that loomed menacingly above, I was overcome by wonder. Spires of steel rose atop a platform undergirded by massively articulated corner pillars. A neo-gothic cathedral? Who are its faithful? Who are its agnostics? Are there heretics among them? What is this religion's architectural worldview, its systemic beliefs?

My guide was Virgil, a spritely but ancient-looking fellow who's known as a scholar of "New Age Religions." The Church of Carbon is one of his specialties.

We disembarked and made our way up the steel staircase to the platform level. As I coped with spiritual vertigo, Virgil informed me that this structure is called a "jack-up" cathedral.[1] The corner pillars are jacked up from religious bedrock located several hundred feet below the ocean's surface.

Virgil dryly observed that the Church of Carbon has many places of worship and devotion. There are not only jack-up oil platforms like this one, but land-based structures such as coal mines, oil refineries, and gas pipelines. And then there are those little chapels called "gas stations" spread out across the land, ministering to the faithful.

As we explored the jack-up monolith, Virgil described it as a prime architectural example of the western Carbon Church sect. There are eastern sects as well. Besides architectural differences, the western sect worships a holy trinity of coal, oil, and natural gas mediated by gigantic corporations.

Virgil vividly described the Church's Nine Pillars of Faith symbolized by the sprawling architecture. The Nine Pillars are divided into upper and lower orders of religious practice. There are five Spires topside and four Footings below. He first explained the symbolic superstructure above, which I summarize as follows:

SPIRES

Limbo – deliberate agnosticism about the relationship between fossil fuel consumption and its impact on the earth's climate. The primary ordinal is a subconscious fear of lifestyle change.

Lust – love of materialism. This pillar is rampant consumerism, replete with fetishes ranging from high frequent-mileage flyer accounts to behemoth SUV's.

Gluttony – waste, that is, energy. Gluttony is accompanied by profound walling off, a separation from nature. Over-cooled, energy-guzzling air conditioned spaces are emblematic.

Greed – stealing from present and future generations. Anointed by crude oil, this tragedy of the commons represents a kind of anesthetic numbness and lack of empathy for those who are, or will be, impacted by climate change.

Wrath – extreme anger and abuse practiced by climate deniers. These self-appointed individuals spew out a toxic liturgy proffered to them by others.

Continuing our architectural tour below deck, Virgil described the Church's foundational elements:

FOOTINGS

Heresy – public actions by people who do know better, but who purposefully create doubts about climate change. These folks are sometimes called "merchants of doubt."

Violence – character assassination through organized *ad hominem* attacks on climate scientists, or on the scientific community at large, as an attempt to discredit the evidence of global warming-caused climate change.

Fraud – toxic liturgy prepared by high priests and used by other church functionaries to persuade the faithful that their future well-being and economic salvation comes only from continued reliance on fossil fuel energy. The word "pollution" is banished from their scriptures.

Treachery – perversion of humans' grant of earthly stewardship in favor of abject exploitation. Propagating the false notion that humans are an exceptional species, immune to the laws of nature and the consequences of dumping vast quantities of carbon into the earth's atmosphere.

Virgil spread his old, spindly arms wide and summed up with a flourish. "This new-age religion," he says, "is all about worshipping the god of carbon at the high altar of the GDP [Gross Domestic Product]."

When he was a bit younger, Virgil had been impressed by German philosopher and biblical scholar Friedrich Schleiermacher's (1768-1834) definition of religious feeling as absolute dependence.[2] "The Church of Carbon devotees' feelings are nothing less than absolute fossil-fuel dependence."

As we were departing, Virgil abruptly pointed, "Look there. Cracks in the Church's edifice! The collapsing coal industry and erosion by millennial heretics throwing off their yokes of carbon."

Harrumph. The sun is still shining, but somehow I find myself more in communion with Dante's *Inferno*.[3, 4]

–Alighieri

4

Leap of Faith

It seems like such a little thing. You've mustered up the courage to broach your concerns about the climate with some friends. But an element of emotional vertigo holds you back, keeping you from jumping into your conversation with them.

Blue Skies – "the quintessential skydiving greeting and salutation. It's like the word *aloha* in Hawaii; it can be used both in greeting and in departure. I think of it as a benediction. It is both practical (yes, we want the skies of blue so we can jump, jump, jump) and spiritual. It is a sacred phrase. Use it often and mean it."[1]

This is it, the blue-skies spirituality that you've been yearning for. Deep down there's a gnawing sense that you've been a total whuffo on the climate's degradation. The only way out is to take a leap . . . a leap of faith. So you're going to "blue-sky it."

You dive out of your comfort zone. The parachute spills and the canopy fills. You're astonished to see a richly textured set of canopy cells or "panels" supporting you from up above.[2]

One panel is the Buddhist Declaration on Climate Change . . . "As Thich Nhat Hanh has said, 'We are here to awaken from the illusion of our separateness. We need to wake up and realize that the Earth is our mother as well as our home.'"

Another panel: the Statement on Climate Change by the Primates of the Anglican Communion . . . "We encourage all Anglicans to recognize that global climatic change is real and that we are contributing to the despoiling of creation. We underline the increasing urgency of this . . ."

Yet another panel is the Baptist Declaration on Environment and Climate Change . . ."We must care about environmental and climate issues because of our love for God . . . This is not our world, it is God's. Therefore, any damage we do to this world is an offense against God Himself."

Panel: Climate Change—An Eastern Orthodox Christian Perspective. . . "The world is charged with the grandeur of God!" . . . "To commit a crime against the natural world is a sin. . . . It is not too late. God's world has incredible healing powers. Within a single generation, we could steer the earth toward our children's future."

Panel: The Episcopal document "Seek Environmental Justice": "Resolved: [to] resist the development and expansion of ever more unconventional, dangerous, and environmentally destructive sources of fossil fuel and move toward conversion to more sustainable sources [and to] support the implementation of grassroots, community-based solutions to climate change."

Panel: The Book of Resolutions of the United Methodist Church: "Be it further resolved that members should make an effort to learn about human production and release of greenhouse gasses and evaluate their own lifestyles to identify areas where reductions in production and release of greenhouse gases can be made."

Panel: The Presbyterian Call to Restore the Creation: "Creation cries out in this time of ecological crisis: . . . Greenhouse gases pose the threat of global warming. The Church has powerful reasons for engagement in restoring God's creation. . . . God's works in creation are too wonderful, too ancient, too beautiful, too good to be desecrated."

Panel: The Society of Friends—Quaker Earthcare Witness. "We recognize the moral duty to cherish Creation for future generations . . . We recognize the connections between climate change and global economic injustice as well as unprecedented levels of consumption . . ."

Panel: The Hindu Declaration on Climate Change: "Hindus encourage the world to be prepared to respond with compassion to such calamitous challenges as population displacement, food and water shortage, catastrophic weather and rampant disease. . . . Mother Earth is no longer in a period of climate change, but in climate crisis. . ."

Panel: The International Conference on Muslim Action on Climate Change: "We believe that global good environmental governance can be achieved with the principles of *mizan* (balance), *khalifah* (leadership of the people and stewardship of the earth), and *ukhuwah* (collaborate, care, and share). These are key elements that . . . promote activities to deal with climate change."

Panel: Jewish Community Priorities for Climate and Energy Policy. "We believe humans remain a partner in Creation. We fulfill the mandate by practicing *Tikkun Olam,* literally, repairing the world. Climate change threatens to irreparably alter the Earth. . ."

Panel: The United Church of Christ's document "Climate Change and Global Warming," quoting the Reverend Dr. Martin Luther King, Jr.: "We are now faced with the fact, my friends, that tomorrow is today. We are confronted with the fierce urgency of now. In this unfolding conundrum of life and history, there is such a thing as being too late."

Panel: The Roman Catholic Church. Pope Francis has stated, "Safeguard Creation. Because if we destroy Creation, Creation will destroy us. Never forget this!"

These strong panels offer strong support to overcome the turbulence of denial and the wind shear of skepticism, with indestructible moral and spiritual shroud lines that attach you to a splendid canopy riding high above.

Skydiving is an act of personal accountability. More, your leap of faith in seeking this accountability is very different from continuing as a fellow traveler, dozing on the plane of apathy. For it will set you along the arc of climate accountability to your children and climate accountability to your God.

Blue Skies!

5

Climate Awakening

At the peak of his career, Ray resigned his position as an E.R. physician to devote himself to fighting climate change. Ever since he was a little kid, he'd always dreamed of being an M.D. Smart, inquisitive, and determined to succeed, Ray always carried deep within him a compassion for people, their travails, and their misfortunes. His action came as a total surprise to all those who know him.

Calm under pressure, deliberate in his ways, strong yet precise, "Dr. Ray" had a knack for "lowering the temperature" in any emergency situation. He was enormously respected by his peers. A natural, they said. And so everyone was puzzled by his seemingly abrupt decision to leave, to forego his life's hands-on work helping people in exchange for something as intangible and controversial as climate change.

At his departure party, which was quite a gala event, Ray made an eloquent speech about the hospital, his colleagues, and the workers up and down the line, all of whom meant so much to him. He also reflected on some of the lives he and the staff had miraculously saved and some funny unexpected things in happenstance. Everyone was feeling good, which was Ray's way. At the end somebody asked Ray why he was taking this drastic step in his life, perhaps in hopes he might change his mind.

Ray's curt response was out of character. He said simply, "You know, this climate thing has gotten under my skin. Sometimes in life you have to trust your heart and go against the grain." And that was that. Dr. Ray had opened up a new chapter in his life. He never looked back.

What was going on with Ray? To his friends, Ray was absolutely the last person they would have imagined becoming a "climate activist." He seemed totally conventional, button-down mainstream. Ray's putting trust in his heart gives us a clue.

There are increasing numbers of people who are putting their normal lives aside to battle climate change. Like Ray, they have come to realize that this threat is not over the horizon, it's happening right now. Once they pass through the gate of climate reality, they enter a world of parallel existence where others—in Ray's case, the good folks at the hospital—go about their daily "climate change" inert lives. Weird!

Ray could no longer abide what George Marshall calls the negotiated social silence on climate change.[1, 2] He had had a spiritual transformation.

We don't know Ray's religious affiliation or even if he's religious at all. But, clearly he had experienced a heartache-induced transformation. All the great religions speak of such transformations. As G.K. Chesterton once wrote: "The church is a house with a hundred gates; and no two men enter at exactly the same angle." The phenomenon has also been studied by academics:[3]

> A spiritual transformation is a profound change in the self.
>
> The change is not simply a matter of maturation, but is typically identified with a process (sudden or

> gradual) by which the transformed self is achieved.
>
> This change in the self is radical in its consequences—indicated by such things as a new centering of concern, interest, and behaviors.
>
> This new sense of self is perceived as "higher" or as emancipation from a previous dilemma or predicament.
>
> This process occurs within a social medium or context. Specifically, spiritual transformation entails a religious framework within which the transformed self is described, acts, and is recognized by others.
>
> Last, if behaviors from these new ideals and changed habits of life do not follow, then there has been no transformation.

Ray simply couldn't stand by and do nothing about climate change. He had passed his gate angling against the grain.[4]

Dr. Ray's heart underwent transformative climate awakening—nothing he could have accomplished in the E.R.

What about you?

6

"God Will Save Us"

Perched on the mesa cliff overlooking a scrub arroyo, it suddenly all came together for Gareth. He had been tormenting himself about climate change. And, whereas some seek science for answers, Gareth looked to God for answers.

When Gareth brings up climate change with his friends in his Christian fellowship group, they say, "God will save us. Humans are too feeble to change the earth's climate. Anyway, even if that weren't the case, God would never let it happen. He loves us too much. End of story."

But Gareth is dubious. He's prone to some altogether different perspectives.

For one thing, Gareth plies his trade as a welder, working on high rises sprouting up all over the Southwest. In late afternoons and early evenings—his own time—he jogs the trail and free-climbs a shortcut up to his refuge. His mesa, his home, his rock, his gig.

There he reads scripture, meditates, and prays. As he does so, the sun slips down over the horizon, ushering in starlight's tight embrace. He stretches his frame across a sun-baked rock.

Gareth knows the climate is changing. It has become obvious to him in the decade he's lived in the Southwest. The blazing heat is becoming more oppressive with each passing year.

When they were riding a steel beam one day, his crew boss dryly observed. "Gareth, you know that big pickup of yours? If you'd point that rig straight up and drive the same speed you hightail out of here after work – well, your sky would be black in 40 minutes. You'd be in outer space!"

Gareth couldn't get this friendly jab out of his head. He Googled some NASA earth images on his Galaxy.[1] Sure enough. In the silhouetted Earth, backlit by the sun, he could just make out the wispy-thin atmospheric layer enveloping our planet. Crew boss was right.

Seeing the earth's image on edge, Gareth was shocked to learn that heat from global warming is equivalent to 345,000 Hiroshima A-bombs blasting the atmosphere every day.[2] "That's real earth on edge."

After months of praying to God about what He makes of all this, Gareth got his answer.

Climate Backstory: Gareth's Parable

According to the Book of Genesis, God created lights in the firmament, "the greater one" that we know as the sun, and then "lesser ones" known to us as stars in the cosmos. And he saw that it was good.

What modern humans know is that God steeped the sun and the stars with hydrogen and other elements which, under the great pressure and confines of gravity, heat up and become furnaces powered by fusion. Whence comes the light and heat warming the earth.

But there is only so much hydrogen in the sun and over the eons it is consumed. It's the same for any star. Paradoxically, as the sun and other stars age, they don't get dimmer, they shine ever brighter. Since God created earth, the sun has brightened over thirty percent![3]

The ever-brightening sun meant that at some point, life on earth would become unsustainable. That would have marked the end for the swarms of living creatures in the waters, the birds in the sky, the creeping things, the cattle and other animals of every kind, and the humans that God created. Too hot.

However, God had a plan for keeping the sun's ever-increasing brightening at bay. He created a natural carbon sink that gradually removes carbon from the atmosphere and stores it in the ocean waters and deep underground as the sun gradually brightens. Today we call that ancient carbon stored underground "hydrocarbons."

God's carbon sink leaves just enough carbon dioxide in the atmosphere to grow life and keep the earth temperature right. Some would say that it's a miracle that the earth has remained habitable over time. But to Gareth, this is simply part of God's overarching plan.

According to Genesis, God granted us dominion over the earth. Some people have taken this to mean that if people rule the earth, they may exploit its resources in any way that suits their fancy.

So they built great civilizations by extracting, mining, and burning the hydrocarbons that God had stored away to guard against the ever-brightening sun. Many humans, including many of Gareth's Christian friends, claim that this is good. Why? Because their lives are good. But what does God say?

Disrupting the carbon cycle isn't in accord with God's will. His grant of dominion is like a proprietorship. According to the Bible, God owns the earth and every living thing on it. Our role is to live as benign stewards, caring for His creations, acting responsibly toward the Owner.

Now that we are on the way to doubling the amount of carbon in the earth's atmosphere, people are noticing the heatwaves, droughts, diseases, and rising tides. Oops! Some are now expecting God's grace. "God will save us."

This is where Gareth realized that they are right. We will be saved though our enabling partnership with God – the old-fashioned grace described in the Old Testament.

According to Gareth the Welder, we must bond and fabricate God's grace. We need to work it into our modern lives to gain our salvation. "And do you know what grace that is?" Gareth asks.

"His gift of abundant, affordable clean energy."

7

The Climate Bystander Syndrome

Two researchers conducted experiments in the late 1960s to observe subjects' reactions in a mocked-up waiting room when increasing levels of smoke were introduced from a hole in the wall. Darley and Latané reported that three quarters of people who were alone in the room reported smoke.[1] But when three people were put in the room together, only one out of twenty-four reported smoke. What was holding them back?

Instead of smoke drifting into a waiting room, smoke from massive fossil fuel combustion is pouring into our earth's atmosphere. A recent Pew Research Center survey shows that nearly half of Americans (48%) consider global climate change a major threat.[2] More than 61% of Americans agree that the earth is warming, but Americans overall "do not rank global warming as among the top threats facing the country." We seem to be stuck as bystanders in a waiting room even though what is at stake is hugely consequential to our nation, our civilization, and the world. Effective action on climate change ultimately depends on our collective will. What's holding us back?

Failure of people to act in social situations—when there are compelling reasons for doing so—is classified by experts as the "Bystander Syndrome." Its roots go back to the parable of the Good Samaritan in the Bible and perhaps to folklore from ancient cultures before. There is abundant evidence that people are reluctant to break ranks from their social groups for fear of disapproval or rejection. The more people in a group, the greater individual reluctance there is to act. In the biblical parable, a Good Samaritan came to the aid of the injured man he encountered along a road when others ignored him.[3]

A fascinating finding from research into the Bystander Syndrome is "pluralistic ignorance."[4] Pluralistic ignorance describes the majority of people in a group who reject a popular sentiment (like climate change not being of concern) but incorrectly assume that most of the others accept it, and therefore go along with the crowd (and continue denying it). In the smoke-filled room experiment, pluralistic ignorance prevented twenty-three of twenty-four people from reporting the obvious. The 48% who consider climate change to be a real threat are bystanding in their rooms, unaware of peers in almost any social gathering who agree with them.

But the aura of pluralistic ignorance is fragile. Just a little exemplary behavior and careful nudging by climate-change Good Samaritans will go a long way in tipping the balance from mass bystanding to action.

Are you one?

Part II

ACTIONS

You've picked up this book and read it. Why did you do it? Answering this question—really answering it—is the first step along the path towards climate action.

This book is meant for people who are regularly in touch with their souls. People who love and respect Creation. People like you.

Your help is urgently needed because the time for climate agnosticism is long past. Climate change is happening all around us. Worse yet, it is accelerating.

What I'm asking you to do is to become a "climate guardian." For this you are supremely qualified. Climate guardianship isn't a commitment to me or any specific organization; it's simply an enduring moral commitment to act on behalf of the Earth's climate.

You see, climate change has been miscast—first as a science problem, then as an environmental problem, then as an economic one, then as a technological one, then as a political one. Yet first and foremost, it is fundamentally a moral problem. And our collective failure to act is really a failure of the human spirit.

Your climate guardianship would be a special form of climate activism to help address this spiritual failure. It takes its lead from the last chapter in this book, "The Climate Bystander Syndrome." The essence of this advocacy is "witnessing" for the earth's climate in order to break through the social conspiracy of silence.

To help you along the way towards impactful climate advocacy, I offer these suggestions:

Read. Start by reading Pope Francis's Encyclical "*Laudato Si*: On Care for Our Common Home." In addition, read George Marshall's "Don't Even Think About It—Why Our Minds are Wired to Ignore Climate Change." Another crucial document is Jonathan Haidt's "The Righteous Mind—Why Good People Are Divided by Politics and Religion."[1]

Look to inspirational role models. Among the people worthy of your respect and attention are Rev. Sally Bingham, Katharine Hayhoe, Bob Inglis, Bill McKibben, and many others. Seek out local role models in climate groups such as 350.org, The Climate Reality Project, and Citizens Climate Lobby. [2]

Set an example. Start walking the walk. Know your personal carbon footprint compared to the world's average. Take measures to reduce it. Inform yourself about the local effects of climate change. Be modest and self-effacing about your green achievements.

Burnish your climate witnessing skills. Seek out a trusted friend in your faith community willing to be your sounding board. Practice the art of the nudge. Avoid gloom and doom. Be a good listener. Stay true to your faith.

Witness to others about the climate at least once a week. Seek out secular and non-secular opportunities. Put climate change in play in your routine conversations. Tell your story.

Keep at it! Many people of faith and goodwill are working diligently beside you.

I return to my original question: Why did you read this book? Once you have your answer, the power of your witnessing and its positive effect on others by your climate advocacy will be unassailable.

Being a Climate Guardian isn't a sentimental "new age" love of Creation. Rather, it's stout-hearted, tough love in defense of Creation. Since the time of Creation there has never been such an opportunity for individuals like you to stand for the very best humanity has to offer. It's a great time to be alive!

—David Cain

Notes

The Promise of Climate Stewardship

1 Kelsey Dallas, "A Majority of Christians Believe God Has Called Them to Care for the World." *The Deseret News*, February 11, 2015.

> https://www.deseretnews.com/article/865621588/A-majority-of-Christians-believe-God-has-called-them-to-save-the-world.html

2 Daniel Cox, Ph.D. and Joanna Piacenza, "Majority of Catholics for Religious Stewardship, One-Third Say God Gave Humans Dominion Over Earth." PRRI.org, June 2015.

> https://www.prri.org/spotlight/majority-of-catholics-for-religious-stewardship-one-third-say-god-gave-humans-dominion/

Note that this article compares Catholics' views with the views of various other religious affiliations.

3 Genesis 1:28, King James Version, in BibleGateway.com.

> https://www.biblegateway.com/passage/?search=Genesis+1%3A28&version=KJV

An Unoriginal Sin

1 Elizabeth Bruening, "Why Do Evangelicals Like James Inhofe Believe that Only God Can Cause Climate Change?" *The New Republic*, January 29, 2015.

> https://newrepublic.com/article/120889/evangelical-james-inhofe-says-only-god-can-cause-climate-change

2 Western culture, particularly in countries like United States, is strongly shaped by Christian religious heritage. This inherited culture continues to influence contemporary values even in the face of growing secularism and increasing cultural diversity. And so, it seems appropriate to refract climate change through the lens of a Christian perspective to gain a more complete understanding. One of the key refractive indexes is the Christian characterization of "sin."

https://www.britannica.com/topic/sin-religion

3 Patrick Leary. "'Original Sin,' Slavery, and American Innocence" in *Social Text Journal,* April 2017.

https://socialtextjournal.org/original-sin-slavery-and-american-innocence/

4 Jeremy Williams, "Sinning against the Earth?" in Make Wealth History online, September 4, 2008.

https://makewealthhistory.org/2008/09/04/sinning-against-the-earth/

Church of Carbon

1 "Jackup Rig," in Wikipedia.

https://en.wikipedia.org/wiki/Jackup_rig

2 "The Restoration of God-consciousness in the Person and Work of Jesus of Nazareth - Friedrich Schleiermacher's Conception of Man, Sin, and the Redemption of Humanity by Christ." Evangelical Catholicity Press online.

https://evangelicalcatholicity.files.wordpress.com/2008/06/schleiermacher-paper.pdf

3 "Canto XXVII. Or, Where Would Dante Put Climate Change Deniers?" Eowyn9, in *Daily Kos,* August 2014.

https://www.dailykos.com/stories/2014/8/4/1319018/-Canto-XXVII-1-Or-Where-Would-Dante-Put-Climate-Change-Deniers

4 "Plot Overview: *Inferno*" in Sparknotes online.

http://www.sparknotes.com/poetry/inferno/summary.html

Leap of Faith

1 "Skydiving Slang Glossary," in Skydive Addiction online.

http://skydiveaddiction.com/blog/skydiving-slang-glossary/

2 "Faith-Based Statements on Climate Change," Compilation by Citizen Climate Lobby Volunteers, July 2015.

https://issuu.com/citizensclimatelobby/docs/faith-based_statments/

Climate Awakening

1 George Marshall,"Why Our Brains Are Wired to Ignore Climate Change and What to Do about It." *The Guardian* online.

https://www.theguardian.com/commentisfree/2014/sep/23/why-our-brains-wired-ignore-climate-change-united-nations

2 "Don't Even Think About It – Why Our Brains Are Wired to Ignore Climate Change," George Marshall, Bloomsbury Publications, August 2014,

https://www.goodreads.com/quotes/278590-the-church-is-a-house-with-a-hundred-gates-and

3 Arthur J. Schwartz, "The Nature of Spiritual Transformation – A Review of the Literature." Published by Metanexus, 2000.

http://www.metanexus.net/archive/spiritualtransformationresearch/research/pdf/STSRP-LiteratureReview2-7.PDF

4 Garth Brooks, "Against The Grain" (song lyrics).

http://www.lyricsfreak.com/g/garth+brooks/against+the+grain_20058030.html

"God Will Save Us"

1 NASA photo of earth's atmosphere in profile, https://www.nasa.gov/multimedia/imagegallery/image_feature_152 9.html

2 Skeptical Science - Getting Skeptical about Global Warming Skepticism: https://www.skepticalscience.com/4-Hiroshima- bombs-worth-of-heat-per-second.html

3 Astronomy Workshop, "Did you know that the Sun was once only 70% as luminous as it is today?", https://janus.astro.umd.edu/front/pages/links/Sunslife1.html

The Climate Bystander Syndrome

1 Latane, B., & Darley, "Bystander 'Apathy,'" *Journal of the American Scientist*, 1969, 57, 244-268.

http://faculty.babson.edu/krollag/org_site/soc_psych/latane_bystand.html

2 Cary Funk and Brian Kennedy, "The Politics of Climate," Pew Research Center, October 2016.

http://www.pewinternet.org/2016/10/04/public-views-on-climate-change-and-climate-scientists/

3 The Parable of the Good Samaritan, Bible Gateway, Luke 10:25-37

https://www.biblegateway.com/passage/?search=Luke+10%3A25-37

4 "Pluralistic Ignorance," Reed College online.

http://www.reed.edu/psychology/pluralisticignorance/

Actions

1 Regarding Pope Francis's encyclical *"Laudato Si:* On Care of Our Common Home":

https://laudatosi.com/watch

Regarding Jonathan Haidt:

http://righteousmind.com

2 Regarding Rev. Sally Bingham:

https://climateone.org/people/rev-sally-bingham

Regarding Katharine Hayhoe:

https://en.wikipedia.org/wiki/Katharine_Hayhoe

Regarding Bob Inglis:

https://en.wikipedia.org/wiki/Bob_Inglis

Regarding Bill McKibben:

https://en.wikipedia.org/wiki/Bill_McKibben

Regarding some organizations worth exploring:

350.org

https://350.org/

The Climate Reality Project:

https://www.climaterealityproject.org/

Citizens Climate Lobby:

https://citizensclimatelobby.org/

Acknowledgments

Special thanks to my brother, Tom, my climate alter ego, who continues to inspire me; to my unendingly supportive and understanding wife, Jan; and to my publisher, Ed Myers, who shares a passion for our climate and whose guidance has been instrumental in shaping this book.

About the Author

David Cain is a member of an extended family descended from a grandmother who was the daughter of Utah pioneers and a Michigan grandfather, who, while working as U.S. government attorney, litigated on behalf of native Americans. David has a Ph.D. in electrical engineering and spent most of his career engaged in R&D for the energy industry. In the early 1990's he co-founded a successful company that provides pollution accounting software for large industrial enterprises. Concerned about global warming and climate change, he spent a year of investigation before retiring to devote himself full-time to climate advocacy. David's ClimateGuardians blog addresses issues of climate change on Facebook's community website.

Visit the ClimateGuardian blog at this URL:

https://www.facebook.com/
ClimateGuardians-1570768463162469/

About Montemayor Press

Montemayor Press is an independent publisher of literature for adults and children. To learn more about our books, visit:

www.MontemayorPress.com

or write for a catalogue at:

Montemayor Press
P. O. Box 546
Montpelier, VT 05601

CPSIA information can be obtained
at www.ICGtesting.com
Printed in the USA
FSHW04n0030040418
46291FS

9 781932 727340